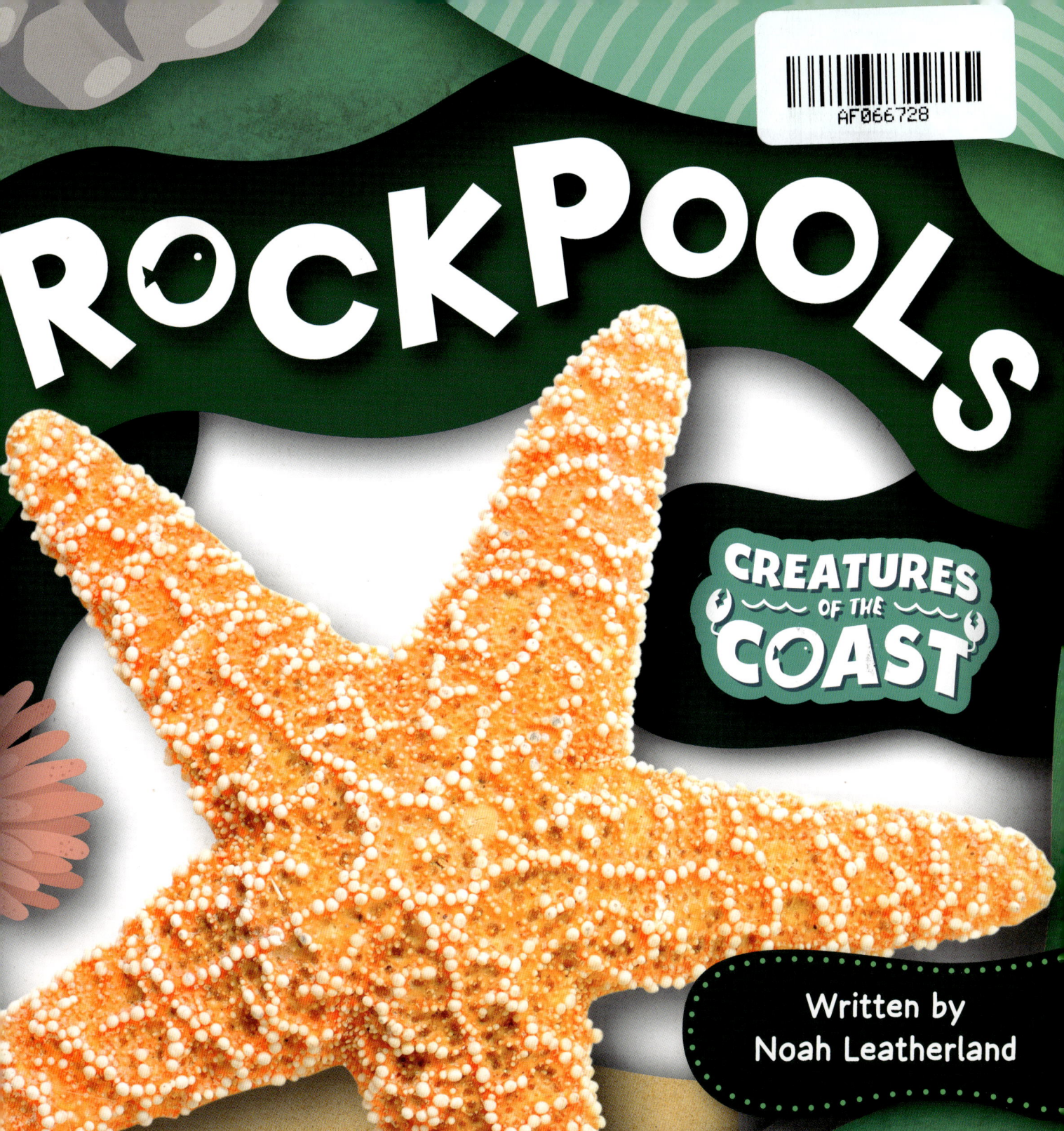

BookLife PUBLISHING

©2023
BookLife Publishing Ltd.
King's Lynn, Norfolk
PE30 4LS, UK

All rights reserved.
Printed in China.

A catalogue record for this book is available from the British Library.

HB ISBN: 978-1-80505-349-1
PB ISBN: 978-1-80505-389-7

Written by:
Noah Leatherland

Edited by:
Rebecca Phillips-Bartlett

Designed by:
Amelia Harris

All facts, statistics, web addresses and URLs in this book were verified as valid and accurate at time of writing. No responsibility for any changes to external websites or references can be accepted by either the author or publisher.

PHOTO CREDITS: All images courtesy of Shutterstock. With thanks to Getty Images, Thinkstock Photo and iStockphoto.
Recurring images: Little_Monster_2070, Perfect_kebab, LadadikArt, Your Local Llamacorn, Olga_Serova, Terdpong, MR. BUDDEE WIANGNGORN, Gabor Ruszkai, Luria, GoodStudio. Cover – aaltair, Groomee, Lexi Claus. 2 – Cary Kalscheuer. 4–5 – FiledIMAGE, Andreas Tychon, avh_vectors. 6–7 – Vara I, Dan Bagur, SewCreamStudio. 8–9 – HotFlash, Karoline Cullen. 10–11 – M.Style, Twinkles503, Fernando M. Elkspera. 12–13 – Anatolir, barmalini, Anthony King Nature. 14–15 – BlueRingMedia, Tomek Friedrich, Pam Walker, Ronald Shimek. 16–17 – Kazakova Maryia, Triple H Images, Keith 316, JPT-SRC, CC BY-SA 4.0 via Wikimedia Commons. 18–19 – Penny Hicks, valda butterworth. 20–21 – Amilat, Alexisaj. 22–23 – Cary Kalscheuer, Juriah Mosin.

CONTENTS

Page 4	In the Rockpool
Page 6	Starfish
Page 8	Sea Urchins
Page 10	Gobies
Page 12	Mussels
Page 14	Barnacles
Page 16	Limpets
Page 18	Periwinkles
Page 20	Anemones
Page 22	Creatures of the Coast
Page 24	Glossary and Index

Words that look like this can be found in the glossary on page 24.

IN THE ROCKPOOL

The coast is the place where the land meets the sea. With rocks, sand and seawater, the coast can make lots of interesting habitats for different animals.

The sea comes in and covers the beach, then goes out again with the tide. Seawater left in the gaps between rocks makes rockpools. Rockpools are home to many amazing creatures.

STARFISH

TUBE FEET

Starfish have long arms that make them look like stars. These arms have lots of tiny tube feet that help starfish stick to rocks. They can slowly crawl along the rocks using their arms.

Starfish are predators. They eat clams, oysters and snails. Starfish eat in an unusual way. They catch a creature, then push their stomachs out of their mouths onto their prey to eat it.

Did you know that starfish are not actually fish?

SEA URCHINS

SPINES

Sea urchins are covered in spiky spines. Sea urchins use these spines to grab onto rocks. Their spines also protect sea urchins from predators. Be careful not to touch them.

Sea urchins have very tiny, hard teeth. Sea urchins use their teeth to eat and to carve holes into rocks. Sea urchins make their homes in these holes.

SOME SEA URCHINS CAN LIVE FOR AROUND 200 YEARS.

Gobies

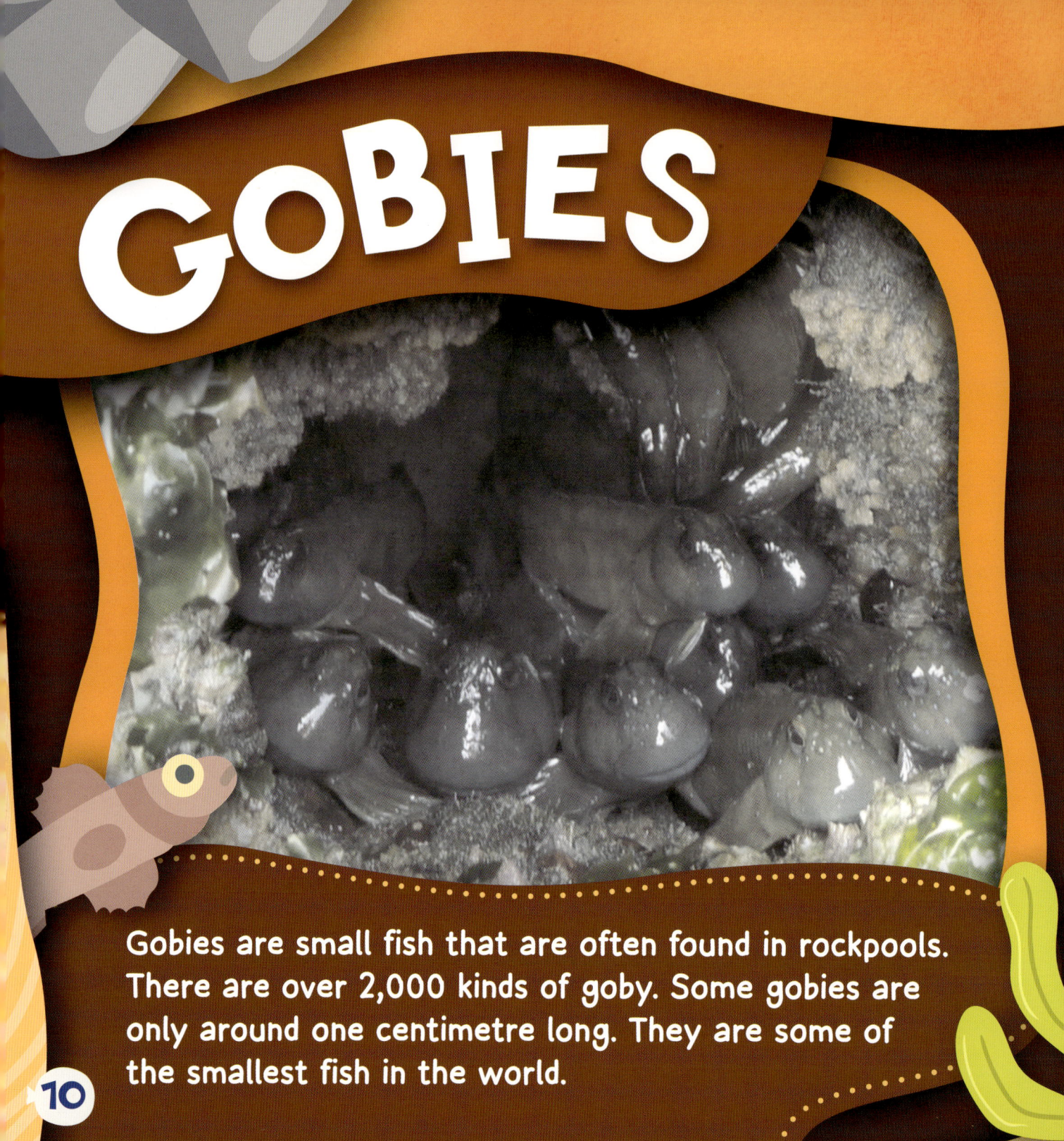

Gobies are small fish that are often found in rockpools. There are over 2,000 kinds of goby. Some gobies are only around one centimetre long. They are some of the smallest fish in the world.

Gobies have adapted to life in rockpools. They can use their fins to stick to rocks so that they do not get washed away by waves. Some gobies can change colour to hide from predators.

FINS

MUSSELS

Mussels are small creatures that live inside their shells. Their shells come in two halves. Birds and other predators open mussel shells to eat the creatures inside.

Mussels are covered in tiny, sticky hairs. They use these hairs to stick to rocks and to each other to make mussel beds. Mussels also use these hairs to trap predators such as sea snails.

MUSSEL BED

BARNACLES

BARNACLES GROWING ON A CRAB'S SHELL

Barnacles are small creatures that never leave their shells. Barnacles can grow in lots of places near water, and even on other sea creatures. Sometimes they can be found in rockpools.

Barnacles do not move once they have attached to something. To gather food, barnacles open their shells and stick out feathery combs, called cirri. The cirri catch tiny creatures, a bit like a net.

CIRRI

Limpets

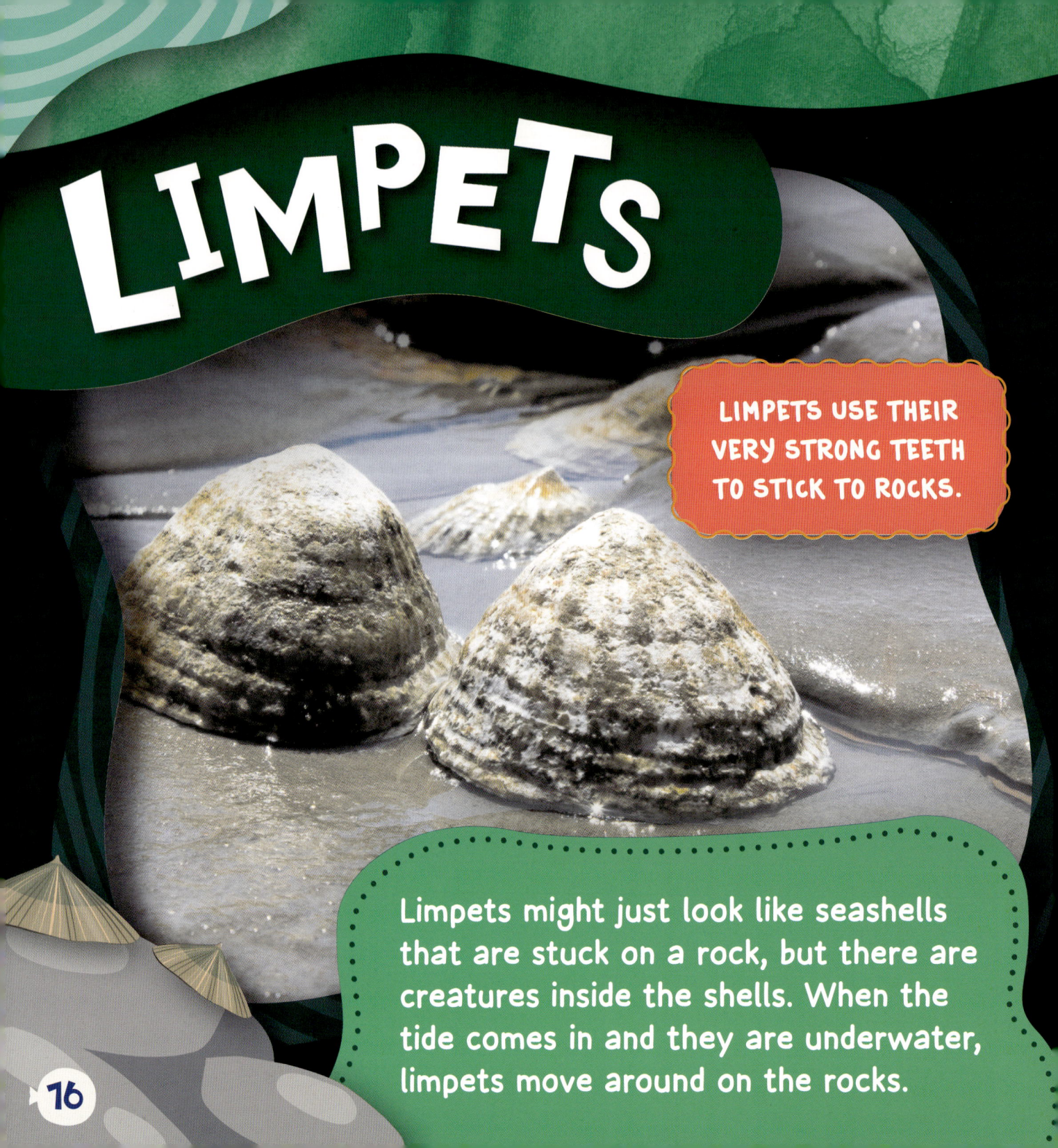

LIMPETS USE THEIR VERY STRONG TEETH TO STICK TO ROCKS.

Limpets might just look like seashells that are stuck on a rock, but there are creatures inside the shells. When the tide comes in and they are underwater, limpets move around on the rocks.

Limpets always return to the same spot on a rock. The edges of their shell wear this spot down and leave a scar. The scar helps the limpet cling onto the rock better.

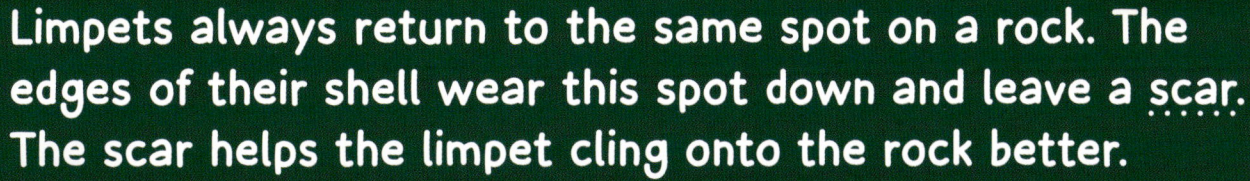

A SCAR LEFT BY A LIMPET

PERIWINKLES

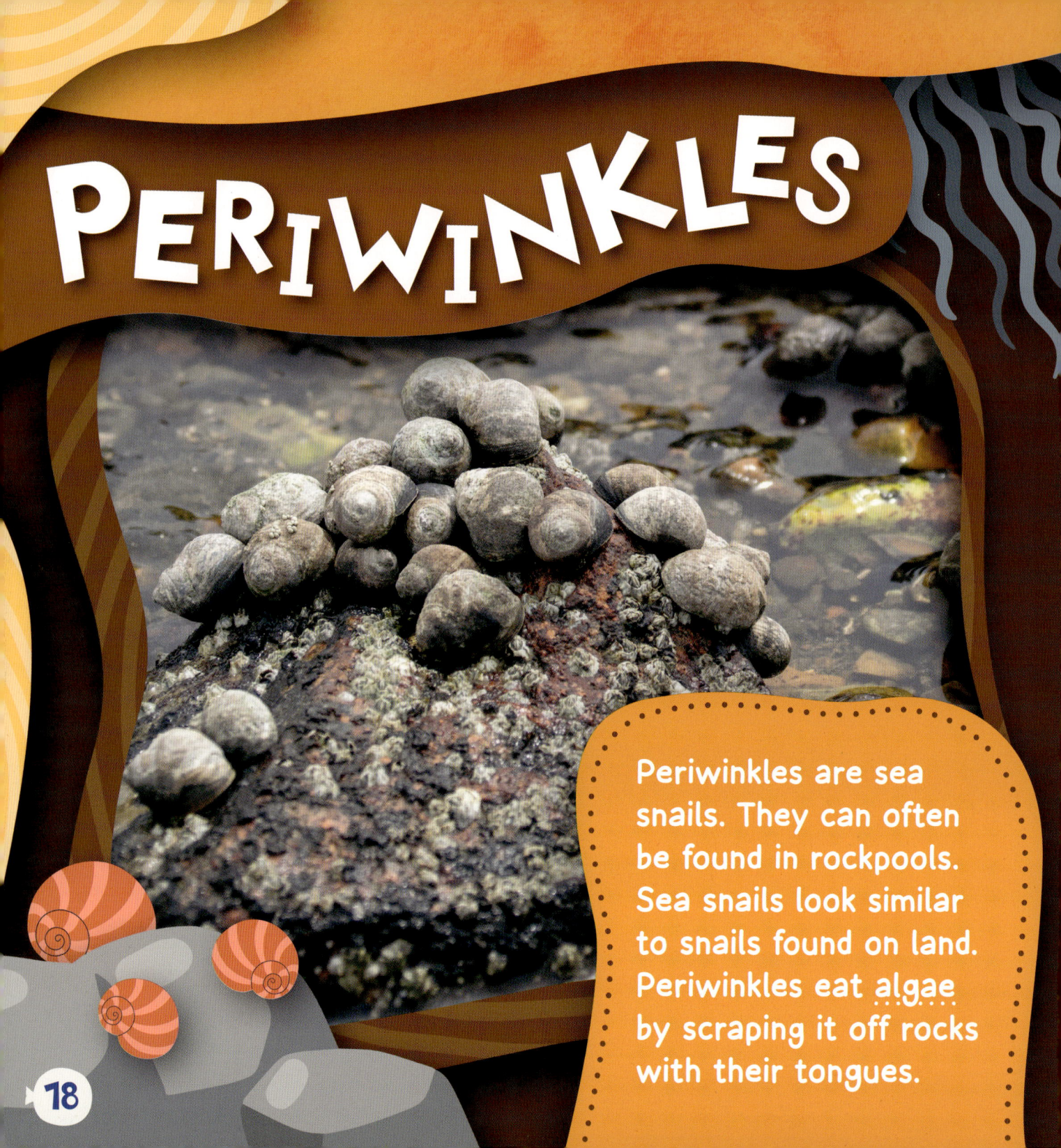

Periwinkles are sea snails. They can often be found in rockpools. Sea snails look similar to snails found on land. Periwinkles eat algae by scraping it off rocks with their tongues.

Periwinkles can tuck themselves inside their shells. Their bodies have hard discs that can close their shells. This helps keep the periwinkle <u>moist</u> when they are out of the water.

ANEMONES

Beadlet anemones can be found in rockpools. They stick to the side of rocks and can move across them very slowly. When the tide is out, they look like small blobs stuck to the rocks.

When the tide comes in, beadlet anemones spread out their tentacles. Beadlet anemones are venomous. They use their tentacles to sting their prey with venom before eating it.

TENTACLES

CREATURES OF THE COAST

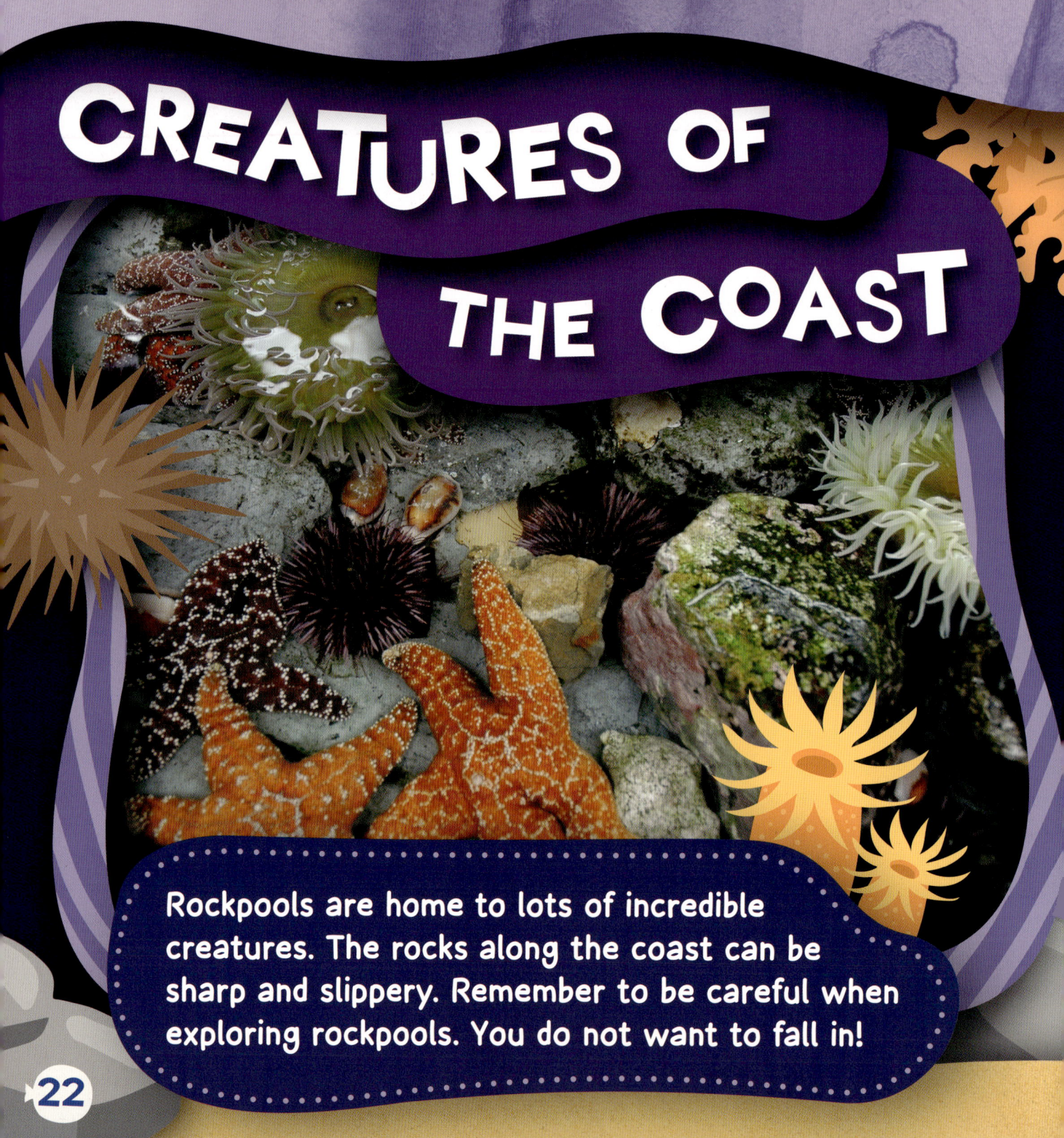

Rockpools are home to lots of incredible creatures. The rocks along the coast can be sharp and slippery. Remember to be careful when exploring rockpools. You do not want to fall in!

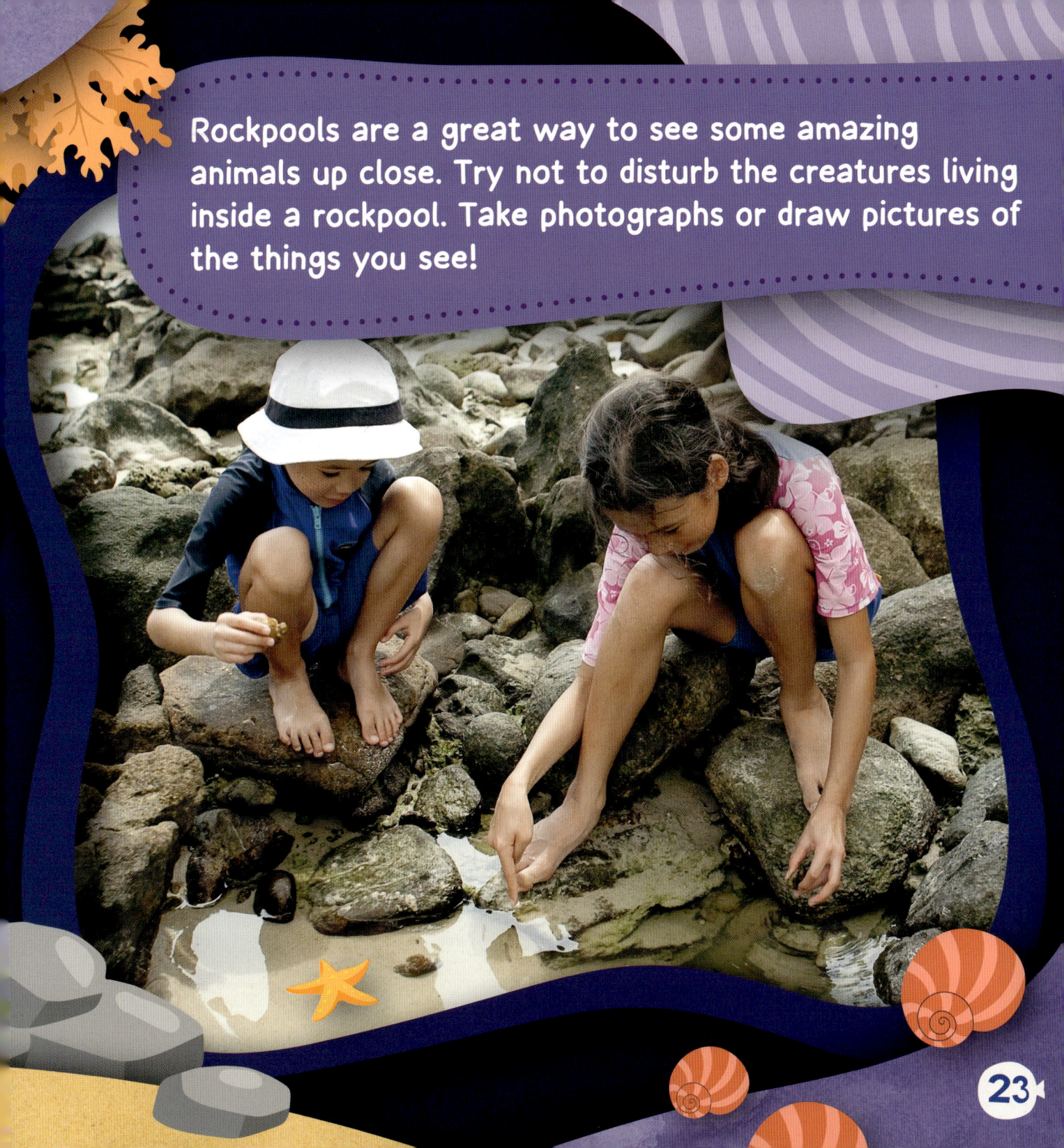

Rockpools are a great way to see some amazing animals up close. Try not to disturb the creatures living inside a rockpool. Take photographs or draw pictures of the things you see!

GLOSSARY

adapted	changed over time to suit the environment
algae	a plant or plant-like living thing that has no roots, stems, leaves or flowers
carve	to cut a shape out of something
habitats	the natural homes in which animals, plants and other living things live
moist	slightly wet
predators	animals that hunt other animals for food
prey	animals that are hunted by other animals for food
scar	a mark left by something that lasts forever
tide	the movement of the ocean towards and away from land
venomous	able to poison another animal through a bite, scratch or sting

algae 18
fins 11
fish 7, 10
food 15
predators 7-8, 11-13
shells 12, 14-17, 19
teeth 9, 16
tentacles 21
tides 5, 16, 20-21
waves 11